Animal Cures
The Country Way

ANIMAL
CURES

The Country Way

Robin Page

DAVIS-POYNTER

London

This book is a collection of cures and remedies used
in bygone days by country people. Although every
care has been taken in recording them, no support is
given for the claims made or implied in them.

First published in 1979 by
Davis-Poynter Limited
20 Garrick Street London WC2E 9BJ

ISBN 0 7067 0240 9

Photoset in Great Britain by
Baird Harris Filmsetting Ltd

and printed by
R. J. Acford, Chichester, Sussex

Designed by Paul Minns
with woodcuts from Thomas Bewick and his pupils

Contents

Acknowledgements

WHILE WRITING THIS BOOK I received invaluable help from numerous people; correspondents, vets, owners of old books, and country men and women who were prepared to talk and to tell me about the 'old days'. There are far too many to name individually, but I am grateful to all of them for their help, consideration and kindness. In addition I would like to thank Patrick Coulson for his pharmaceutical guidance, and to Teresa Brown, who, in my absence, managed to type out the final draft, from an almost illegible manuscript.

Introduction

ALTHOUGH MOST PEOPLE are familiar with several of the old country cures for humans, such as the use of snails for getting rid of warts, and the chamber-pot to combat chilblains, few have any knowledge of the vast number of old animal cures that were once tried for domestic pets and farm animals. Yet on farms and in old books, a great medicine cupboard of cures and would-be cures still exist: they remain in the memories of country people, and written down on scraps of paper, damp and discoloured with age, but many continue to be used with success.

Now, of course, few people are in everyday contact with animals, apart from dogs and cats, and when they become ill they are whisked away to the veterinary surgeon who dishes out both the necessary pill and the appropriate bill. In the past, however, things were very different, and through the ages, as man plodded along the road towards supposed 'civilisation', he took animals with him, and in many cases he depended on them. The earliest hunters used dogs to help them take their quarry, and then cattle and horses provided food and muscle power. But as the animals worked for man, so a close bond of affection developed between them, and the dog and horse in particular, became his close companions.

Even up to the Second World War, many people were involved with animals; milking cows, leading horses at plough, feeding the pigs and hens at the bottom of the garden, and most homes had cats and dogs. The importance of good animal care is shown by the old saying:

Let poverty set foot in the stable,
And it will soon be all over the house.

When animals became ill it was often a matter of great urgency, both financial and physical, and action was taken using simple and easily obtainable remedies. My grandmother, a strict teetotaller, would ferment sloe wine to feed calves if they developed scour, and peppercorns were pushed down the throats of day old chicks to prevent them from catching cold. Father too can remember eggs, shell and all, being given to calves to aid their digestion, and old Jim, who used to work on the farm, was a great believer in eucalyptus to get rid of fleas on dogs.

Most of the cures were straightforward and based on common sense, although my local vet says that several of them contain sound veterinary principles. Some too involved buying drugs and herbs from the chemist, among them substances with fine sounding names such as Tartarised Antimony, Peruvian Bark, Powdered Catechu and Tincture of Rhubarb. They were measured by the drachm (an eighth of an ounce), and in ounces. Most of the powders can still be obtained and used today, and according to one old farm-worker, they should be mixed into potions and given with 'either a quart of beer or human urine, if you can tell the difference these days'. Another liquid measurement often used was the 'gill', a quarter of a pint.

The purpose of this collection of cures is not to replace the modern-day vet, for with innoculations against hard pad in dogs and cures for milk fever in cows, they prevent much suffering; but there is also no doubt that many animals are fussed quite needlessly when afflicted with common ailments, and they could be treated at home. Consequently this book is meant to be a simple practical guide, using the old ways, although I have again included some cures for amusement only, and these should be obvious. As a practical guide, I am also aware that many remedies have been omitted; indeed, a record of all the old cures, traditions and superstitions associated with animals, would fill many large volumes.

1 Animals in the Home

THE MOST COMMON ANIMALS in the home are dogs and cats. We have always had them, and being on a farm they live almost idyllic lives; the dogs are well fed, they have plenty of exercise in the fields and along the hedgerows, and when there is nothing else for them to do they go ratting, or lie in the sun. The cats too are comfortable, spending their time sleeping in a warm nook among the hay bales, or hunting mice. Living in this way they generally enjoy excellent health, with the cars on the roads being their greatest threat. Other dogs and cats are not so lucky, and those owners who do not have large gardens or live in the country should ensure that their animals get plenty of exercise, for inactivity and overfeeding cause much illness and disease.

DOGS

Of all the domesticated animals, dogs probably have the longest association with man, and many people agree with the old saying that: 'A dog is a man's best friend.' Some owners, particularly in the past, became so friendly with their dogs that they actually went to sleep with them in front of the fire, for warmth and companionship, leading to the observation:

He that lieth down with the dogs, shall rise up with fleas.

This need not be true, however, for if a dog is active it often loses its wildlife, or at least it does not share them around by scratching, for:

> *A dog that is idle barks at his fleas,*
> *But he that is hunting feels them not.*

If a dog does have fleas, they are simple to kill or disperse. A single flea is easily caught; spit should be applied to a finger, and then the finger to the flea, and the flea will stick to the saliva. Even a 'lousy' dog can be cleared quickly; a gypsy remedy is to put the dog on a lead and swim it backwards and forwards in a river or stream.

When it comes out, the fleas will make for the surface of the coat, but they will be too waterlogged to jump; they can then be wiped off with a hand or a cloth. In the North of England 'lousy' sheep-dogs are plunged into sheep dips, while Jeyes fluid is another liquid that fleas find intolerable. Tansy leaves mixed with straw keep them out of the kennel, and flea and louse powder can be made from 'louseberries', the fruits of the spindle tree. They should be baked and powdered before use, and the powder can also be used on people. If eucalyptus is tried, it should be applied outside, otherwise the fleas will jump all over the house to get away from their strong smelling host.

Worms, or segments of worms, are other parasites that can be brought indoors. Dogs and cats pick them up from the rats, mice and rabbits they catch, and they are not very pleasant. Raw carrot is a straightforward remedy, as are pumpkin seeds, whole or ground. Garlic, or a dessertspoonful of mustard in two tablespoonfuls of milk, can also be effective. One old writer praises the Areca nut, which comes from Malaya, while another claims that walnut leaves are just as good and can be obtained more easily: 'In summer, when the leaves are green, they must be dried and baked in a plate

before the fire, then rubbed to a fine powder with the hands.' Two heaped teaspoonfuls should be mixed in half a pint of warm milk and given for eight consecutive days. The fact that caterpillars do not eat walnut leaves is given as proof of their effectiveness.

However, apart from having a few unwelcome guests, not only is the dog man's best friend, but it can also be his faithful defender. Consequently, when approaching an unknown dog in somebody's garden, it is worth taking note of its bark, for 'A dog will bark ere he bite.' It is also said that: 'Dogs that bark at a distance bite not the hand.' If a dog does eventually bite, then it is reassuring to know that:

A dog's bark is worse than its bite.

The ones to approach with the greatest caution are those that watch, or growl softly, hence the warning:

3

Beware of a silent dog and still water

and

Beware of the silent man and the dog that does not bark.

When an aggressive dog is encountered there are several ways to deter it. First of all it should be stared at menacingly, eye to eye, which can make it feel less secure, and money in a pocket should be rattled loudly. In a real emergency the threatened person should get on all fours, put a cap or a paper in his mouth and shake his head. In theory, the dog will then back away.

Normally the relationship between man and dog is good however, and the easiest way to judge a dog's health is to touch its nose, as:

A dog's nose and a maid's knee are always cold.

The nose should also be damp, so if a dog has a warm dry nose it is probably off colour. On occasions the dog will seek its own remedy by eating couch-grass or 'twitch', which is also known as 'dog's grass'.

A dull coat is another indication of illness, although shine can sometimes be restored by adding sea-weed to the diet. In 1814, Colonel George Hanger, a noted sportsman, had a more unusual tonic; he wrote: 'When a dog looks unkindly in his coat – then give him three doses of powdered glass, as much as will lie heaped upon a shilling to each dose.' He claimed that it worked, but he did not advocate the use of any old glass: 'The powdered glass must not be made of the green glass bottles, but from broken decanters and wine glasses, powdered and ground in an iron mortar, then sifted through a fine muslin sieve.' Some people also use ground glass for worms, which could in fact be the cause of the coat's poor condition.

At one time bad coats, and even baldness, were common in dogs and resulted from mange, which is caused by mites in the fur. According to a retired farmworker: 'In the old days, dogs often had the mange; they went around half naked. We used to treat them with salt and water, or warm water and Dettol. An old man would say "That old dog's got the manger." That's where we get the saying "dog in the manger." '* Another mange remedy involves garlic, or elderleaves, simmered in a quart of water and then rubbed into the coat; if lemon juice is added, the mixture will combat ring-worm as well. Mange can also be treated if four ounces of flowers of sulphur, two ounces of oil of tar (obtained by distilling wood), and two drachms of carbolic acid are added to a pound of melted lard and stirred until cool.

The dog should be washed, using Castile, or another good soap, and then dried thoroughly before the mixture is rubbed over its whole body. The process should be repeated two or three times in a fortnight. One old remedy advocates the use of 'train oil', but with the reduction of railway lines in favour of motorways, it is no longer practical.

Diet is very important in maintaining a dog's health and Colonel Hanger believed that it was porridge oats that gave sheep-dogs their stamina and resistance to cold. An even earlier sage recommended the delicacy of 'boiled cow noses' with Scottish or Irish oatmeal. Chopped parsley in the food will aid general health, and it also helps to rectify anaemia. Purges were another favourite way of retaining good condition, and they can still be used; aloes and Glauber's salts were particularly popular. Aloes comes from the leaves of several species of the aloe plant, found in hot climates, and Glauber's salts is the popular name for sodium sulphate, which can be

---

*His interpretation of 'dog in a manger' is wrong – it comes from the dog actually being in a manger, and although not eating the hay itself, preventing the horses or cows from having it. Consequently a 'dog in a manger' is somebody who prevents others from having what he doesn't want.

used to reduce temperature and as an antidote to carbolic acid poisoning.

An active dog will eat more than its owner provides, for it may develop a round of friendly households for bones, and raid the occasional dustbin. If wire or glass is swallowed as a result, bread and milk must be fed in large quantities, or cotton wool should be forced down the throat, to put a protective covering around the sharp edges. Inevitably too, being naturally inquisitive and greedy, many country dogs eat poison put down for slugs and rats. Force feeding a lump of washing soda will help, and a mixture of mustard and vinegar will cause immediate sickness to evacuate the stomach. Being rather unselective eaters, dogs will also eat things like cooked asparagus stalks and even old, smelly baler twine. If they do, constipation will result, and apple cores, blackberries and strawberries will help to ease the condition. In the summer, most dogs snap at passing bees and wasps, and I once had a fox cub that actually ate them: if they get stung, particularly on the nose, washing soda or a piece of cucumber should be applied.

All energetic dogs are also liable to injure themselves. Sore pads can be treated with a lotion of hog's lard, (or Vaseline), and potato peel or ivy leaves, and a bath in Friar's Balsam (Tincture of Benzoin), will help torn pads. The paws and pads of gun dogs should be washed with strong salt and water in cold weather, and warm water and soap when the ground is hard and dry. If a sprain results, a paste of whitewash and cowdung is the answer, and it can also be used on horses.

Deep wounds can be treated by covering them with cobwebs, puff-balls or a lotion made from rosemary leaves, or ground ivy, in hog's lard. The leaves should be boiled up in the lard and then strained off. When the lotion cools it sets, and can be applied with the fingers. For smaller cuts and abrasions, water can be used, after blackberry or elder leaves have been boiled in it. Sometimes, if the cuts and scratches result from fighting, abscesses or infection can result; if these do occur, a turnip, parsnip or swede poultice will bring relief.

Eyes too are vulnerable, and again an old gypsy claims to have the answer; he uses an eyewash made by boiling ground ivy in water, and a lotion of ground ivy in lard will cure most dog skin troubles, including eczema. If ground ivy is not available for the wash, fennel, elder blossom or chickweed can all be used, and cucumber juice squeezed into the eye is another way to give relief.

Canker is one of the commonest ailments; it irritates the ears and causes the infected animal much discomfort. A very old cure suggests that the dog should be rolled over onto its side so that a lump of soap, as big as a walnut, and a tablespoon of brandy can be put into its ear. The ear should then be closed and rubbed until the soap begins to lather. The rubbing should continue until three or four tablespoonfuls of brandy have been used, and the soap and brandy are well mixed. Similar treatment should be given for three to four days. For those who do not like seeing their brandy disappearing down a dog's ear, there are cheaper and simpler methods. Warm olive oil, chamomile lotion, or two teaspoonfuls of lemon juice in two dessert-spoonfuls of warm water are soothing and effective.

Anybody with a bitch will know that her periods of 'heat' can be a trying time, with all the dogs of the area fighting and cocking their legs around the garden. Their passion can seldom be stifled, even buckets of cold water and gunshots make little impression. The answer, therefore, is to hide the scent of the bitch; this can be done by pouring paraffin everywhere she squats, but unfortunately with some bitches it would be difficult to carry enough paraffin. If the problem is the exact opposite, then balm, thyme, marjoram, fennel, or wild water mint can be tried as aphrodisiacs.

Should the paraffin treatment fail, or the water mint succeed, the birth of the puppies can be eased by occasionally feeding some raspberry leaves during the gestation period, whereas three or four ivy leaves stirred in a mug of hot water

7

will get rid of the afterbirth. Honey in milk is very good for puppies, and when they are cutting their teeth, cabbage stalks will save chewed up slippers and furniture. The vet should be contacted to get an early innoculation for distemper, and failing this, leeches, should be applied to the ears.

Old dogs need special care and they should not be allowed to get cold after exercise. If colds and coughs do occur, they can be eased by giving blackberry or blackcurrant jam with honey. In the event of pneumonia setting in, the chest should be rubbed every hour with a strong mixture of turpentine and mustard. Rheumatics can be helped by feeding raw parsley or cooked nettles, and for arthritis the joints should be massaged with olive oil. Cataracts need a liquid made from the flowers and leaves of the greater celandine, and violets eaten, and applied externally, have sometimes been known to relieve tumours.

Rabies is one of the most dangerous diseases in dogs, and it is almost impossible to cure. When people caught it in the middle ages, after being bitten or scratched by a 'mad' dog or cat, they were sometimes suffocated by relatives to avoid suffering. Infected animals froth at the mouth, drink plenty of water, and eat their own excreta. Colonel Hanger advised that if a pet was bitten in the head by a 'mad dog', it should be destroyed immediately, but if bitten only slightly, it should be chained up, and the bite cut out and burnt with a hot iron. Liquid caustic soda should then be poured in and rubbed, two or three times a day.

Of course, just as there are old diseases and dangers, so there are many modern hazards. If a dog suffers from shock after being hit by a car, it must be kept in the dark, quiet and warm, and given honey and brandy. Scalds and burns should be treated with honey, vinegar or a raw potato, and if bones are broken, chopped comfrey should be put in the food; cabbage stalks can be used instead of plaster of Paris.

An old gamekeeper once claimed that he had the cure for virtually all dog ailments. He would put a lump of washing soap as big as a walnut on the back of the dog's tongue and make it swallow. His small grandson would then take the dog for half an hour's walk, before returning it to the owner and collecting half an ounce of Robin Redbreast tobacco as payment. The grown up grandson now says: 'The treatment gave them the runs, but it seemed to cure them.'

The best treatment for a dog, however, is simple, and is contained in the old proverb:

> *A good dog deserves a good bone.*

CATS

Cats and dogs often appear to live quite contentedly in the same house, with the same owner, but it is misleading, for:

> *The cat and dog may kiss, yet are none the better friends.*

This is confirmed by the fact that if given the chance, the average dog will chase a cat, even if it has been curled up with it in front of the fire two minutes before.

There are fewer country cures for cats than for dogs, for the usual problem was too many cats, with numerous families of unwanted kittens appearing each year. This led to many strays and hungry cats:

> *A cat is hungry when a crust contents her.*

In addition many people considered that whereas the dog genuinely showed affection for its master, the cat merely wanted a warm place to sleep and eat, and used man to its own advantage. Hence in one eighteenth-century book it is recorded that: 'The cat is assiduous to please, but it is sly, distrustful, and treacherous. She will take advantage of your inatention to steal your breakfast.' Cats also seem to be

9

independent and well able to fend for themselves, and this is aided by their cleanliness and resilience. If they fall from a great height, they invariable land on their feet, and the saying that: 'A cat has nine lives', has been well won.

If a new cat is wanted it should be recalled that:

Wanton kittens make sober cats.

and

Never was a mewing cat a good mouser.

Inevitably, an efficient mouser will get fleas and worms. The fleas can be lost by boiling a lemon, complete with rind, and applying the liquid to the cat's head and ears. Ticks too are sometimes encountered, they can be made to release their hold by burning them with a lighted cigarette or a smouldering twig. For worms the cat should be made to swallow a wad of tobacco.

Because of their teeth and claws, it may not always be possible to treat a cat with tobacco or anything else. To avoid bites and scratches when treating or inspecting a cat, it should

be rolled up in a sack or a thick cloth. A good way of deceiving a cat into taking tasteless powders, or crushed pills, is to mix them up with margarine or dripping. The mixture should then be smeared on to the cat's paws and muzzle; when the cat licks itself clean, it will dose itself at the same time.

'That cat is out of kind that sweet milk will not lap' and therefore if a cat is ill or sickening for something, it will eat couch-grass and cat mint. Diarrhoea can be treated with crushed clay pipes, a simple way of getting kaolin, and liquid paraffin (the medicinal oil, not the fuel), will cure constipation. Both can be mixed with cod liver oil, or a sardine, for easy consumption. If it gulps the sardine, bicarbonate of soda will cure hiccoughs.

'Caterwauling' at night indicates that the cat is love-sick and nothing more. Females can breed after six months, and the smell of cat mint will arouse them sexually. Cuts and bruises obtained in the resulting fights can be treated with plantain, and to prevent a cat from scratching raw places, the long sharp part of the claw can be clipped, as it is nerveless and bloodless. If a cat attacks sores with its mouth, or scratches its ears, a lamp shade should be placed over its head and fixed at the neck.

When a cat sneezes the easiest treatment is to turn it out of the house, for an old belief suggests that the whole household will otherwise catch cold. It is also widely held that cats have a premonition of death and will creep away to die quietly in a secluded place.

At one time it was thought to be unlucky to move with a cat into a new house, and on a change of address the poor animal was simply left behind. Cats can be helped to settle into a new home by covering their paws with butter, to make them sit down to wash, and by putting them in front of the kitchen oven for the night. If a cat does stray it can be enticed back with the powdered roots of valerian, and some find it so intoxicating that they will roll about in the plants.

2 Animals in the Farmyard

AT ONE TIME a variety of animals and fowls could be found in many cottage gardens, or on common land, and consequently most country people knew their ways and recognised their ailments. But over the years things have changed, with more people preferring poodles or flowering shrubs outside their back doors, to pigs, and even on the farms themselves the old farmyard scene of sows rootling and hens scrapping are less familiar. However, there are signs that moods are changing and that some people want to get back to simple things, consequently those wanting 'self-sufficiency', or just fresh eggs, are keeping hens, and Jacob's sheep are in great demand as mammalian lawn-mowers and suppliers of wool. Perhaps too the pig will regain its popular place, for once styes stood behind many homes, and the inmates were used to 're-cycle' waste, long before that phrase had become fashionable or had got any popular meaning.

PIGS

Until comparatively recently, the backyard pig was common,

12

and it converted all edible household waste into pork. Hence the old saying: 'Swine are the chief Support of the Kitchen; and on the other Hand, the Kitchen is a great Support to the Swine'. Another old proverb says: 'The worft Houfwife will reare the beft Pigs', which was, and still is, true, for the worst housewives have the most dirt and waste. The farmer too was expected to have his own pig for 'The Farmer is improvident and unwife, who feeks his Bacon at the Butcher's'.

Because pigs were so useful, successful breeding was important, and to ensure good litters, a bryony root was often

hung up in the sow's sty, and some country people continue to proclaim its virtue. The bryony was used, because it ressembles the Biblical 'mandrake', an aphrodisiac and a symbol of fertility.

According to a sixteenth-century belief, a sow will have as many piglets as she has teats, for 'Every sow has juft fo many Paps as she brings Pigs at a Fare (litter)' and 'Every Pig knows his own Pap, and fucks at that and no other'. Thay also believed that each piglet would go to the 'Pap' that was closest to it in the womb, and that if a pig was taken away, the milk from the allocated teat would dry up. However, simple observation does not always confirm these beliefs.

When piglets are born to a young sow, their backs should be smeared with a mixture of ink aloes and water, to prevent their mother from eating them, and the sow should be given plenty of barley made soft in water. If scouring starts, one ounce of Robin's starch should be made into a paste, added to a gill of scalded milk, and given once a day for four days: the treatment can also be given to calves.

Piglets grow quickly, and it seems to be true that:

A pretty pig makes an ugly sow.

The pig's curly tail is said to be a sign of good health, but like so many animals, even healthy pigs can get worms and lice. The downy inside of hips, the fruit of the wild rose, gets rid of worms in the small intestine, and if cinders and coal are fed they will get rid of them elsewhere. A pig with a good muddy wallow will lose its lice and ticks, for when it rubs itself against a tree, they will be rubbed off with the mud.

If and when illness does strike, however, it should be treated immediately. For measles, the sick animal should be starved for three days and nights and then fed on five or six apples each day, on the assumption that a pound of apples a day keeps the vetinerary surgeon away. The apple cores should be cut out and the holes stuffed with flowers of brimstone (sulphur); the treatment should continue for five or six days. Measles can be avoided by putting bryony root and cummins seeds in the water and by tipping the food into lead troughs, although there is no record of a remedy for lead poisoning. Pox sores can be treated with two large spoonfuls of treacle in a pint of warm water and honey, taken internally, and a mixture of flowers of brimstone, tobacco dust, and hog's lard, applied externally.

When a pig gets bitten by a viper or a 'mad dog', the wound must be washed out with human urine or warm vinegar, and Black Hellebores (Christmas Roses), should be put into its ears. Other lumps and bumps can be treated just as simply: 'quinsy', is a swelling of the glands in the throat, and if they become too large, they should be cut, squeezed, and washed out, before a blend of coarse brown sugar and yellow soap is applied. 'Kernals' is another swelling of the throat, and the infected pig should eat daffodil bulbs, which work in a similar way to onions.

The most unusual cure, which could even make an ugly sow attractive, is for 'Diftemper in lungs': 'Pierce both Ears of the Hog, and into each Orifice, place a Leaf and Stalk, a little bruifed, of the Black Hellebore': again an interesting way of arranging flowers.

SHEEP

Lambs are the favourite farm animals of most people, thanks to nursery rhymes like: 'Mary had a little lamb'. But there are dangers if a tame lamb is kept in the garden, for if it is male: 'A pet lamb makes a cross ram', and many captive males, even deer and badgers, grow up into belligerent and even dangerous adults.

One early naturalist had a great admiration for rams and wrote: 'The ram is strong and fierce – he has even been known, regardless of danger, to engage a bull; and his forehead being much harder than that of any other animal, he seldom fails to conquer; for the bull, by lowering his head, receives the stroke of the ram between his eyes, which usually fells him'.

If a lamb grows into a ewe, it should not be allowed to breed in the first year, and again the old advice is sound: 'If the Shepherd has Regard to the Coupling of the Rams with Ewes, when they are all fuffer'd to go in one flock: to prevent the Rutting of the Young or other Rams with the Ewes that are too young, he will tie Twifted Rufhes, Bits of Leather, or fomewhat of thefe sorts to the Tails of the Ewes he would prevent from Breeding; which method will put the Ram from his intent'.

If 'rutting' is wanted, but it does not occur, then a diet of knot-grass, the blades of onions, and the leaves of turnips, will encourage them into action. Illness at lambing time in the ewes can be overcome by feeding them ivy, and to keep them strong, each one should have a daily ration of one pound of powdered linseed and two pounds of oatmeal, boiled together like porridge; some sugar and one or two tablespoonfuls of gin or brandy can also be added. Young lambs must also be protected from crows and ravens, which will peck out their eyes if they get the opportunity, and if a twin is introduced to a mother that has lost her lamb, the new infant should be rubbed with the

fleece of the dead lamb to acquire an acceptable scent. When castration time comes, the oldest and simplest method is for the shepherd to bite off the required parts with his teeth.

All sheep, whether young or old, should be sheared in early June, as instructed by the saying:

> *You may sheer your sheep,*
> *When the elder blossoms peep.*

Then the animals can be washed and allowed to dry without catching cold, and when the fleeces are off they will remain warm.

Sheep, because of their nature and their wool, get several

unpleasant parasites and ailments. 'The fly' is when maggots hatch and start to eat the animal alive; they can be treated with a strong mixture of boiled tobacco, or by chewing tobacco and spitting over them.

Rot is another problem, and:

> *Ewes are more fubject to rot than Rams,*
> *Unlefs it is when they fuckle lambs.*

An old cure for foot-rot:

¼lb ground alum
½lb sugar of lead (now only available in solution)
½lb blue vitriol (copper sulphate)

They should be mixed in a gill of water, applied with a feather, and kept in an air-tight container when not in use.

The other great scourge of sheep is 'scab', caused by mites in the wool; a simple lotion for a few sheep can be made from one pound of tar mixed with half a pound of goosegrease, hog's lard, hen dripping, or unsalted butter. For a whole flock, broom salve can be used; the ingredients are; twenty gallons of spring water from gravelly soil, and ten gallons of the tops, leaves, stalks and flowers of broom. They should be simmered gently until reaching the consistency of jelly, then two quarts of stale human urine should be added, together with two quarts of salty beef or pork brine, and two pounds of melted mutton suet. After a minute or two of stirring, to mix the suet, it should be strained off and used when required. The mixture will also rid the fleece of ticks and lice, as will a quarter of an ounce of tobacco and a drachm of salt mixed with a quart of water. A good diet is also important in combatting scab, and sheep should never be put out to graze if there is mildew on the grass.

Other more general 'diftempers' can be cured with beaten juniper berries mixed with half a bushel of oats (on average

17

about twenty one pounds) and a quarter of a pint of sea salt; scab is rarely found amoung sheep where junipers are growing wild. For a cough, or shortage of breath, the sheep should be 'bled in the ear', before a mixture containing oil of almonds and white wine is poured into their nostrils a spoonful at a time. It is advisable to separate any animals with measles or pox from the rest of the flock, and if the sores and spots are treated with three ounces of rosemary leaves boiled in strong vinegar, relief will be given

An old hand-written remedy for sheep with the 'trembles' or 'moss illness', recommends: 'One gill of warm beer, one table-spoonful of treacle, and one ounce of yeast. Brew your yeast up and then put it into the beer and treacle and then give the full gill. Leave the sheep where it is lying and don't move it at all – in half an hour it will be up and on its feet'.

For sore eyes, celandines are again advised, although as an alternative ground ivy leaves can be chewed up and the juice squirted out of the mouth and into the infected eye. When sheep have worms, or have eaten a horse leech, or a poisonous herb, 'bleed them in the lips and under the tail'. To prevent sheep being worried by dogs, twenty out of every hundred should wear a bell the size of a tea-cup.

GOATS

Goats are very hardy animals and seldom seem to get ill. If they do, many of the cures for sheep and pigs can be used as and when required. Again, ivy leaves are considered to be good for the 'nanny' at 'kidding', and helps to bring the afterbirth away

HENS

To get good hens, fertile eggs are first required, but an old belief suggests that if eggs are carried over running water they will become infertile. A clutch should contain thirteen eggs, for if an even number is set, they will either not hatch or the chickens will all be cockerels. To make a recently broody hen stay on a new clutch, she should be spun round to make her dizzy, and then sat down with a sack over her head. Spring wildflowers

should never be picked and carried into the house when a hen is sitting. To make a broody hen comfortable, ensure that a dry dust bath is available to get rid of lice, and pyrethrum leaves in the nest-box will keep off fleas. If red mite became a problem, a coating of creosote on the nest-boxes and perches will destroy them.

After the chicks have hatched, a peppercorn put down the throat of each one will prevent colds, although gape can be a danger to them. It is caused by small worms in the throat and lungs; it may be treated by stripping a feather of all but its tip, dipping it in eucalyptus, pushing it down the throat and twisting it around. A more comfortable method is to feed them small crumbs of dough impregnated with a little soft soap. When their wings and feathers seem to droop, soap and soot rolled into a ball will ease the condition.

In the summer, hens will often eat long dry grass, and when they do, they run the risk of becoming crop bound. They should be made to drink warm water, before massage is applied to the crop. The head must then be lowered and the crop squeezed, to try and force out the contents. A teaspoonful of Epsom Salts is then required. In bad cases a small hole can be cut in the crop and the grass drawn out. After sowing up, with silk or horse hair, the bird must be fed on soft food to avoid a repetition of the trouble.

'Run the hens and make them lay' is the recommendation once the chickens have become adult, and 'Any old hen lays in March'. Egg shells should not be burnt on the fire, for the hen concerned will start laying lush eggs (eggs with soft shells), or stop laying altogether. When they are layed, the eggs pass through a U-shaped bone, and if it is wide enough to hold three fingers, the bird is a good layer. If it will only take two fingers it is fair, but if it will not take one, the cooking pot is the simplest remedy. Similarly, the best cure for a broken bone is a hot oven and parsley and thyme stuffing. My farming father did once cure a cockerel with a broken leg; unfortunately, when he took the plaster off, he found that he had set the foot backwards. Whenever a small egg is found, without a yoke, it is said to be the effort of a cockerel.

If a hen goes broody when it should still be laying, buckets of water can be thrown over it, but it will remain undeterred. The best way to move it, is to make it sit in front of a cold draught.

When hens moult, the growth of new feathers can be encouraged by feeding dessicated grasshoppers, which are rich in silicic acid in a soluble form. Unfortunately the use of chemical sprays in fields and gardens has meant that a cure is now needed that does not use grasshoppers.

OTHER FARMYARD FOWL

Turkeys are reared on many farms for the traditional Christmas dinner, and boiled greater celandine will rid them of yellow diarrhoea. Black-head, when the birds droop and die because of a small parasite, is rarely encountered if the birds are fed plenty of stinging nettles and lettuce.

'Goosegrass', or cleavers, is very good for newly hatched goslings, while in the old days when adult geese were driven to market, they were first made to walk through warm tar and then loose sand, to get a protective covering for their feet. Ducks can get corns on their feet, and when the ground is hard they should have plenty of grass or straw to walk on. If ducks get digestive troubles, charcoal and ashes will restore them to health. Clutches of duck eggs should be dampened regularly, to soften them for hatching, and May hatches should be avoided; if they do arrive then, ducklings will often pick up a 'bug', roll over onto their backs, and die.

On wet land the possession of ducks can actually prevent disease, for they eat snails which are the intermediate-hosts for liver fluke in sheep and cattle. The fluke settle in the liver and can cause death. Old shepherds often put ducks with their sheep to keep their flocks healthy, and I recently visited a deer farm in a 'fluke' area, where ducks and Canada geese are still used to eat the snails. As a result the deer are completely free from fluke.

3 *The Stable*

FOR HUNDREDS OF YEARS horses have been used for work and for pleasure, and for generations the old cart horses, shires, percherons, clydesdales, and Suffolk punches provided the power on the land. Now the position has changed, with most horses being used for hunting, jumping, racing and riding, and the heavy horses that remain are kept by breweries, or those who remember the past with nostalgia.

But whatever type of horse is kept, the problems relating to them are similar, for all horses need care, warmth and consideration: if they are missing, then the horse's condition will deteriorate, causing danger to itself and loss to its owner. The situation is best summed up by the saying:

> *For want of a nail the shoe was lost,*
> *For want of a shoe the horse was lost,*
> *For want of a horse the rider was lost.*

At one time, as so much depended on the horse, the original purchase was very important; for as one old farrier wrote: 'Next to choosing a wife, buying a horse to carry you "for better or for worse", is the affair in life that requires most deliberate circumspection. True, the former is proverbially a "lottery", but there is no reason that the latter need be! Care should still be taken, for according to Nimrod, the nineteenth-century sporting writer: 'The moment a man has a horse to sell he becomes a suspicious character'; rather like a second-hand car dealer today.

One of the most common ways of cheating with a horse is to tamper with its teeth, for they can be filed, or even knocked out to deceive over age. In addition, broken winded horses can be improved for the duration of a sale by dosing with 'atropine', a drug obtained from deadly nightshade. A more lasting cure, used by old horsekeepers, is the common wildflower herb robert, boiled in beer. Consequently the purchaser should remember that:

You can't judge a horse by its harness.

And another helpful proverb for the would-be buyer is:

A good horse can't be a bad colour.

However, it is also important not to be too critical for:

Who'er expects a perfect horse to see,
Expects what never was, or is, or e'er shall be.

As with all stock, the hazards of purchase can be avoided by breeding. To get a mare and a stallion into an amorous mood, bryony root can be used, while to 'provoke Luft', the mare should drink clarified honey with new milk; her hinder parts should then be touched with a bunch of stinging nettles, and 'she will receive'. When the stallion dismounts, a bucket of cold water should be thrown over the 'mare's part', to help it close up and retain the semen. If the mare has to return to the stallion she should again be touched with a nettle under the tail, and the horsekeeper should say to the stallion: 'Another foal now boy'. Sometimes the mare may object to the advances of her prospective mate and lash out with her back feet. If the stallion gets kicked in the eye the best treatment is an application of goose dung or an eyewash of half a tablespoonful of common salt in a wine bottle full of spring water.

To tell if the mare is 'in foal' a mouthful of water should be squirted into one of her ears. If she is pregnant she will shake just her head, if not, she will shake her whole body. Once pregnant, the gestation period is eleven months, for: 'A mare and a hare go a year'.

An easy birth can be helped by feeding rowan berries (mountain ash), and in the event of a mare not cleansing (losing her afterbirth), she should be given 'a quart of strong beer with a handful of fennel and a fourth part of olive oil, mixed and administered milk warm up the nostril'. When a mare's milk dries up, half a gallon of 'warm ale' as well as top quality hay and grain is usually the answer. If the milk flow does not return, however, and the foal gets a stomach upset from drinking cow's milk, two or three spoonfuls of rhubarb powder, with an equal quantity of magnesia, in warm gruel, will settle the condition. Scour, or 'white chute', can also be treated with one ounce of Robin's starch, made into a thin paste, and mixed with a gill of scalded milk. The mixture should be given once a day for four days, and it can also be used for calves. If the bowel is just irritated, a tallow candle can be inserted up the rectum.

The old horsekeepers of the past really cared for their horses and to get shiny coats they would bake bryony root, powder it on a nutmeg grater, and give two spoonfuls a week. A pinch of arsenic twice a week, plenty of carrots, linseed cake, or a pint of soaked wheat has the same effect. Another method is to apply 'worm oil'; the oil is made by filling a jar with worms, sealing it, and putting it in a muck heap for three weeks. If a horse is being ridden or worked, it is important that it is kept comfortable; elder in the halter will keep flies away, as will a garland of bryony leaves, for most insects find the crushed stalks unpleasant. Walnut leaves boiled in water and then sponged onto the coat will also keep flies at bay. To prevent a sore back, a folded blanket should be put under the saddle, or alternatively, handfuls of 'arsemart' can be used; it is water pepper or smartweed, and was given the name 'arsemart' by an old 'medic': 'because if it touch the taile or other bare skinne, it maketh it smart.'

Care of the legs and feet is very important and if degeneration of the feet occurs the animal is probably suffering from 'thrush'; the affected foot should be covered with cow dung and tied round with sacking as it 'will draw the rubbish out'.

Mudfoot too is an irritation of the feet, often found in muddy conditions, the infected hoof should be cleaned out and daubed with Stockholm tar, before again covering with a sack or cloth. A turnip poultice is the treatment for 'grease', when the foot develops a greasy discharge, and the cause is not fully understood.

For swollen legs the horse should be given one pound of nitre and half a pound of flowers of sulphur made into a mass with honey or molasses, while sore or brittle hooves need an application of linseed oil. At one time lameness and rheumatism in working horses, sheep and cattle, were thought to be caused by shrews running over the limbs. A shrew had to be caught, buried alive in a hole in an ash tree, and then the affected leg or joint had to be rubbed with leaves or twigs from the tree.

It was important too, that a working or hunting horse did not get cold after much effort, and its water intake had to be controlled, hence: 'Let a horse drink when he will, not what he will.' However, if a horse is off colour or awkward, it is also true that: 'You may lead a horse to water but you can't make him drink.' As horses have a 'sweet tooth', if brown sugar, or molasses, is put into water, they will rarely refuse.

In the event of a cart turning over, throwing the horse as well, the horse's head should be sat on immediately, to prevent it struggling and injuring itself. Sometimes a heavy load can lift the horse up in the shafts; when this happens the leather straps of the harness must be slashed with a knife as quickly as possible. If cuts should result they can be treated by applying a mouldy apple, and halves of apples were once commonly left on stable beams for this purpose; oil of turpentine, or water and brandy are other alternatives.

Glanders, or 'farcy', results in nodules growing on various internal organs and can cause much suffering. One of the most effective antidotes is a handful of rue, (a plant from Southern Europe), boiled in a quart of ale and given when warm. Rue juice put into a horse's ears, and kept in with wool is an additional help. More unpleasantness is caused by the bot fly, and when horses stampede around in the summer, the flies are laying eggs in their coats. Later the horse licks the larvae from its coat which allows them to travel through the stomach and intestines, until they pass out nine months later, with the dung, to pupate in the ground. Colonel Hanger

27

did not believe that worms could survive in a horse's stomach and wrote: 'I have often read, in farriers' works, and in those of veterinary surgeons of worms in a horse's stomach: for my own part, I cannot credit it; for the peristaltic motion of the stomach is so powerful and the heat so great, when the horse is alive, that I am of the opinion that worms may well live between mill-stones, when at work, or in a hot baker's oven, as in a horse's stomach;' but the bot flies survive, and can be very troublesome. Groundsel boiled in water will get rid of them, (and cure the staggers), but once the larvae have reached the intestines, boiled linseed oil and tobacco should be syringed up the rectum and held in with a rag, or, like the little Dutch boy when faced with the hole in the dyke.

According to one old remedy, colic and gripes can be cured with a pint of warm beer mixed with an ounce of ground ginger. If the horse won't take its medicine however, its tongue should be pulled to one side and the bottle emptied into its mouth. To make it swallow, its throat should be punched lightly. Another cure for colic can be administered much more easily. A small onion is simply pushed up the rectum of the sufferer and the animal walked around until it improves; it is not known whether this cure makes the horse's eyes water. If a horse will not pass water, it needs a mixture of old beer and vinegar, and its stomach should be hit lightly with elder leaves and twigs.

At one time 'horsemen', or 'toadmen', claimed to have power over horses, by possessing a special bone from the body of a toad.* Although they often understood horses, they also carried out their claims and threats by devious means, for example, by dragging dead rats through a manger to put a horse off its food, and sprinkling pepper over stable doors to keep the inmates inside. If vinegar is rubbed onto a horse's nose, it will behave normally, for it will only smell

---

*See 'Superstition' in *Cures and Remedies: The Country Way.*

vinegar. A saucer of milk at the stable door will also overcome bad smells. Old horsekeepers would often carry a small hare's foot sprinkled with aniseed or rosemary to combat the work of any mischievous toadman, and rowan was used against all evil influences. It was hung up in the stables, lodged in harnesses and hat-bands, and horse whips usually had rowan handles. If a horse is upset by a latter-day toadman, or anybody else, it can be made manageable by adding a pinch or two of ground frog's bone to its food. As frogs are now so scarce, a better alternative is to rub its nose with powdered rue, hemlock or feverfew. Half a drachm of powdered hellebore root, or digitalis (from foxgloves), and half an ounce of liquorice powder, mixed into a ball, with syrup, can also make an effective sedative.

If a horse groans while it works, it does not signify the interference of a 'toadman', it is a sign of loyalty and industry, for:

> A groaning horse and a groaning wife
> never faileth their master yet.

4 The Cowyard

ACCORDING TO AN EARLY WRITER: 'Of all the animals which have been domesticated by man, the cow is unquestionably the most valuable.' At one time cattle provided meat, milk, skins and manure, their blood was fed to pigs, or put into 'muck heaps', their fat was used for candles, their hair went into cement, and their bones and hooves were made into glue. In addition some were used to pull ploughs and carts, and whenever a cowpat landed in a field, a cowslip was said to grow. They are still important for milk, meat and leather and to ensure a healthy herd, horseshoes and stones with holes in them should be hung in the cowyard. In addition, garlands of bryony leaves will keep witches away, and bunches of primroses left on the floor of the cowshed on 1 May, will have the same effect, as witches are particularly active on that day.

BREEDING

To ensure plentiful milk and a continuing herd, calves are required each year, and so again breeding is very important.

If the artificial inseminator is not used, it should be remembered that:

> He that will have his farm full,
> Must keep an old cock and a young bull;

conversely, it is also true that: 'Old kine more milk, young hens more eggs! The bull must be in good health to do his job properly, and before the age of Victorian prudery, one cure was described as: 'Remedy for fwoln cods in a bull! The prescription is, two quarts of strong old beer, a handful of young elder shoots, and two handfuls of the bark from the woody part of a blackberry bush. They should be boiled gently, before bathing the bull's 'parts' with the liquid, night and morning.

Once the bull is fit: 'Cattle of the female gender are said to have copulative instincts excited by eating a small quantity of the plant youthwort or lustwort' (purple loosestrife). A 'hot' concoction, to engender a similar state in the cow, contains four ounces of cummins seeds, one ounce of long pepper, one ounce of ground ginger and half an ounce of cantharides (dried Spanish fly), mixed with three pints of water: 'hunger the cow overnight, then next day give the cow this drink and do not give it any water for two hours and it will come on bulling in a few days time.'

'To make a cow stand at bull, stamp the roots of squills or sea onions (a plant from the Mediterranean), add to water and rub her under the tail therewith! If a small cut is made in the cow's ear after being taken to the bull, it will ensure that she is in calf. If the cut is not made and the cow does not get in calf, she should be given four ounces of cattle salts, one pint of vinegar, one ounce of ground alum, and one teaspoonful of common soda, crushed up very fine: 'Mix this all right well up together – give in one drink one or two days before bulling. But do not let her in with the bull the first time – and give her the same drink on the second time – then

31

let her in with the bull and it will hold.'

Once the cow is pregnant, abortion can be prevented by letting a donkey, or a goat, go with the herd, for it will eat all the harmful weeds without suffering any ill effects itself. A further aid to prevent abortion is two teaspoonfuls of crude carbolic mixed with a pint of warm water. It should be given after the cow has gone over six weeks, and it should be administered three times a week for six weeks. The old piece of paper on which the cure is written says: 'If you give this drink to the cow as it is here stated it is Perfectly safe and it is a sure Prevention for the abortion in cows this has Been used and always with the very Best results obtainable.'

THE MATERNITY SHED

If the first calf of the year is born during daylight, the rest will follow its example, as they will if it is born during the night. Once the calf has arrived, salt should be put on the cow's back, to ensure good health, and the bisnings* given away; the bucket must not be washed, even when empty, or the cow will go dry. Stepping over a calf can cause it to die, and if death does come, the heart should be hung up in the chimney full of the thorns of blackthorn, to prevent a similar occurrence. A cow not 'cleansing' should be given a bucket full of warm water containing one pint of wheatmeal and as many burning wood embers as possible, preferably oak.

At the birth of a calf, 'milk fever' can be very serious. The afflicted cow will fall down, and if not given immediate help it may die. A vet can usually ensure quick recovery by injecting a pint of calcium, but if he is not available and

---

*Known in some areas as 'beestings'. It is the thick milk produced straight after calving. It is popular with country people and makes delicious custards. Officially it is not allowed to be sold as it is considered to be 'unfit for human consumption.'

prevention is required, there are several alternatives: four or five quarts of blood can be bled from the cow eight or ten days before the calf is due; the cow can be milked three times a day before calving; the afterbirth can be hung in a hawthorn tree, or the udder can be blown up with a bicycle pump.

The calf too can be in danger, and 'scour' is very common. When it occurs the calf should be given sloe wine, or cobwebs mixed with a whole egg; a variation of this is to get the calf to swallow an unbroken egg. If acorns or string are eaten, two or three tablespoonfuls of melted lard and a small drink from a bucket will start things moving. Many farmers consider lard to be more effective than castor oil.

Hoose worm in the lungs, 'husk', can also be a problem; the calf should be given one dessertspoonful of turpentine and three tablespoonfuls of linseed oil. An hour afterwards a small teaspoonful of paraffin oil should be poured down one nostril. The next morning the drink should be given again and the oil poured down the other nostril. After four mornings the hoose will go.

A calf should be weaned when the moon is waning, and 'a lowing cow soon forgets her calf.' When the horns begin to appear they should be rubbed with caustic soda or burnt with a hot iron, to prevent them from growing, and when a bull calf is castrated, its scrotum should be sown up and annointed with fresh butter.

THE FIELD AND THE COWSHED

It is an old belief that once a cow is in milk, buttercups rubbed onto the udder will improve the quality. A swollen udder is best cured with one pound of goose grease whipped into a white paste with half a pint of spring water and then rubbed in. Chickweed, groundsel or plantain, boiled in a gallon of whey can also be used; a piece of soaked flannel should be put onto the affected part, as hot as possible, and repeated every three hours. If mastitis develops, the udder should be massaged with dock or elder leaves.

During the spring and summer, there is a danger that cattle can become 'blown.' The condition is often caused when they are put onto new grass, or when they get into other crops. According to one old guide it is caused when: 'they through their greediness to eat, forget to lie down to ruminate or chew their cud. Thus the paunch, or first stomach is rendered incapable of expelling its contents; and a concoction and fermentation takes place.' It can also occur if a cow lays down flat, and it is often an additional complication when a cow is down with milk fever.

When an animal is blown, a cane with a knob on the end should be forced down the throat, to release the air. In extreme cases a knife should be stuck between the haunch and the last rib on the left side. Fetid air will rush out and a small tube should be inserted; when fully deflated the hole should be covered with plaster.

Cows also get digestive troubles and 'flatulent windy colic' is best treated with four ounces of bruised linseed and one ounce of tobacco boiled in three quarts of water. It should be strained through a linen cloth and syringed up the anus when 'new milk warm.' Diarrhoea, scour, or 'the bloody flux', can be brought under control with nutmeg, or a quarter of an ounce of grape or raisin pips, powdered, and boiled in a quart of strong ale. If pips are not available the inner bark of oak can be used as a substitute. One old farmer always uses warm Guinness to cure scour, but unfortunately, the cow only gets alternate bottles; consequently as the animal gets better, he becomes progressively worse. Diarrhoea can also be a symptom of black-water and redwater fevers; hog's dung and nettleroots, pounded and boiled in a blacksmith's anvil water, is the way to cure these complaints.

If ash leaves are hung in a cowshed, snakes will be kept away, but cows can be bitten if they lie down in a grass field to chew the cud. Olive oil boiled with five handfuls of plantain leaves should be applied to bites, for plantain has a good reputation; some old countrymen claim that weasels roll in it before attacking vipers, to give themselves protection. Like their human counterparts, imported Irish cattle are slightly different; if an adder gets on to where an Irish cow has been lying, it cannot get off and dies.

In damp changeable weather, cows can catch cold, and warmth can be provided with old sacks, if any are now available. In addition the sick animal should be given two ounces of ground ginger, a half pound of treacle, and one good ounce of fresh butter, in two quarts of warm beer. If a fever develops, then a handful of cabbage leaves should be boiled in water with a little salt, some butter, and an ounce of treacle. It should be given every morning for four or five days.

Foul foot is common and can be treated with a poultice made from two handfuls each of ragwort and brooklime, pounded to a pulp, and added to wheat flour and water. A simpler method is to watch where the affected foot touches the ground: the imprint should then be dug up and either hung in a blackthorn bush, or turned over – it is said to give a complete cure in five days.

The good cowman can use many more simple cures: herb robert can be used for ulcers and internal haemorrages; the leaves of the greater celandine will cure ringworm, and onions hung around the neck of a sick animal will cure cattle plague; the onions should be changed each day, and the old ones burried in a deep hole. In fact for all contagious diseases, the whole herd should be driven through the smoke of a fire started by rubbing two sticks together. Even the removal of an oat husk or a barley flight can be achieved simply by blowing in white sugar; a teaspoonful of sugar in the eye will also cure New Forest Eye within a week. With horned cattle, cuts and wounds are common, and these can be dealt with by applying black slugs beaten up with black soap. Bran poultices are also good and should be put on as hot as possible. Unfortunately, while

35

trying to tie sacks onto the bran with baler twine, to keep it in place, the cow will buck and kick, and the treatment will often resemble a rodeo.

There is one other condition which cow owners may see. In 1790 James Woodeford, the Suffolk parson, wrote in his diary: 'My poor cow Polly is not able to get up yet as she has a disorder which I never heard of before or any of our Somerset friends. It is called tail-shot, that is a separation of some of the joints of the tail about a foot from the tip of the tail, or rather a slipping of one joint from another. It also makes her teeth quite loose in her head. The cure is to open that part of the tail so slipt lengthways and put in an onion boiled and some salt, and bind it up with some course tape.'

However, in 1723 a Professor Bradley described the same condition as 'the tail'. His remedy was to insert a mixture of 'salt, garlick and wood sutt (soot)'.

5 Wild, Tame and Unwanted

BECAUSE OF HIS EARLIER close connections with nature, man has always kept many wild animals and strange creatures, in addition to those of use to him. Some have been caught and tamed, and others have just been confined for their beauty, or for the novelty of keeping them as living ornaments. Some too have been found injured, and they have been cared for and restored to health, before being released back into the wild, or kept on as members of the family.

Wild animals can still be taken in today, but preferably only if they have been injured or orphaned. The help they need is usually found out by trial and error, and we have had a great variety of birds and beasts over the years; some have become close pets, and others have died causing us much sadness. On one occasion a small black duck from Northern waters arrived on our lawn, after being blown off course in a gale. It was exhausted, but with warmth and plenty of porridge oats it was quickly restored to health.

Foxes have been my particular pleasure, and one of my cubs was cured of constipation when syrup of figs was added to his milk. All my foxes have thrived on household scraps, although

their favourite delicacy has been cooked hens' gizzards. Even now my brother has a young orphaned hedgehog, and that was helped into hibernation with milk and generous helpings of fried and boiled egg. Rabbits too are simple to keep, but if one begins to scour a handful of raspberry or blackberry leaves twice a day will usually cure it in two days. I have never had a badger, but an old country belief suggests that the legs of badgers are shorter on one side than the other, to allow them to run along furrows or the sides of hills. As most houses don't slope, badgers are best left in the wild, and in any case it is illegal to take them.

At one time bears were kept, either as pets or as a means of making money, and if they become fashionable again, the advice contained in an old book should be taken. The writer states that escaped bears can be lured back with raspberry jam, as they are particularly fond of raspberries. The lashings of jam should be deposited quickly, as an escaped bear can be in danger from those who wish to use it as a medicinal aid; for if a bald head is rubbed with a coarse cloth, and then annointed with 'bear's grease', the baldness will be cured.

It is a strange fact that several old men have knowledge of cures for camels. This is not because of any association with wild camels, but because they used them during the First World War. One book of animal management published at that time has a whole chapter devoted to them, with useful tips on 'crossing ditches with camels' and 'marching with camels'. Also, if a camel tends to do the splits, it can be helped by strapping its back legs together. Indeed, a study of camels

can be interesting, and it has recently been discovered that the old herdsman's habit of putting an almond into the female camel's uterus, was one of the earliest contraceptive devices; it worked on the same principle as the modern day coil. If more bizarre pets are required it should be remembered that a hippopotamus can eat as much as six hundred pounds of grass a day, and the black rhinoceros gets its bad temper from its diet of thorns.

Some people however, do have more orthodox pets, and birds in cages and avaries are popular. Gerarde, the old herbalist wrote that chickweed should be given to 'little birds in cages when they loath their meat'. Canaries and budgerigars thrive on rapeseed, groundsel and lettuce during the summer, and eating apples in the winter; they should be given in addition to their normal seed. Syrup of buckthorn will help to get rid of colds and plantain seeds will help clear stomach upsets. If boiled egg is chopped and mixed with soaked bread the mixture will strengthen newly hatched young.

Wild birds should not be caught and put into cages, for not

only is the practice cruel, but many old country people still believe that if a young bird is caught and caged, its parents will try to feed it poisonous weeds through the bars to release it from misery.

If birds have bad eyes, the rich sap of the greater celandine should be applied. At one time falconers used it to improve the sight of their birds, and folk tales suggest that swallows apply it to the eyes of their young in an emergency; in fact the original name for the plant 'chelidonion', comes from the Greek word meaning swallow. If by any chance swallows are seen in walls or puddles, they should be left alone, for according to Cruden's Concordance in 1737, the swallow 'appears in spring and summer, and goes away in autumn. It is thought that it passes the sea, and withdraws into hotter climates, where it either hides itself in holes in the earth, or even in marshes, and under the water, wherein sometimes great lumps of swallows have been fished up, fixed one to another by the claws and beak: and when they are laid in a warm place, they move and recover, though before they seemed to be dead!

Fish, and not swallows, are normally found in water, and many are kept either in bowls or in garden ponds. Lettuce leaves and lean pork, dried and powdered, are good for gold-fish in bowls. If fish are troubled by fungus growth, a crystal of permanganate of potash usually clears the trouble. A sprig of parsley in a bowl, or a handful thrown into a pond will heal sick fish.

Unfortunately, if domestic animals and pets are kept, rats and mice will be attracted. There are several unpleasant old ways of getting rid of them, such as filling their holes with a mixture of cement and broken glass, but less cruel methods are also easy to use. Valerian is a good bait for rats, and it is thought that the Pied Piper of Hamelin stuffed his clothes with the herb in 1284, to make himself irresistible to them. If the attractive lesser celandine is mixed with food it will actually poison them, while crushed bryony root will drive them from their holes. Small pieces of fried sponge, left near the holes, with water available, will also kill them, for on drinking, the sponge will swell up and they will die.

But once the rats and mice have been killed, the problems are not over, as they will sometimes die under floorboards, leading to an unpleasant smell which can be difficult to locate. If this does happen, some blow flies should be caught and put into a milk bottle. They should then be released in the shed or room where the smell permeates; they will soon settle over the dead animal, the appropriate floorboard can be taken up, and the cause of the trouble removed.

6 A Calendar Guide to Animal Health

JANUARY

JANUARY IS NORMALLY one of the coldest months and so warmth is essential. The cat and dog will seek the comfort of the hearth rug and livestock must be given plenty of clean straw.

With cows a close watch should be kept for ringworm and lice. The grass has stopped growing, and any cattle still outside must be given plenty of good hay.

Hens need to be kept dry and a dust bath should be maintained if possible, with plenty of fresh ashes and cinders. Horseshoes and bryony leaves, hung up, will give all the animals an added guarantee of good health.

Snowdrops will appear, and in a mild year the bird's will begin to sing, as a promise of the approach of spring.

FEBRUARY

THE GEESE WILL START to lay: 'For on Valentine's Day, will a good goose lay', and incubation may start before the month is out. Make sure the sitting goose is not worried by foxes. Old geese should also be protected, for: 'A blind goose does not know a fox from a fern'.

There will be days of intense cold and the hearth rug should still be offered to the working dog after a cool wet day. The pads of dogs must also be checked carefully.

Hay should be kept in good condition and all the animals should have as much exercise as possible. Leg and hoof problems can cause difficulties with cows and horses if there is too much mud.

February litters of piglets must be protected from chills and the first lambs will be born.

A definite dawn chorus will confirm the earlier rise of the sun.

MARCH

IN A NORMAL YEAR there will be real periods of spring sun-shine, but frosts and snow can also show that winter has not completely gone.

Lambing will be in full swing and on cold nights the lambs can be helped to their first feed. Plenty of ivy leaves should be available for the lambing ewes and a warm pen kept ready for any orphans. The lambs must be protected from crows, ravens and foxes.

Egg production will increase with the longer days and the dog can spend more time in its kennel. If the soil dries out and land-work resumes, working horses should be watched for sprains and strains. Make sure that their saddles and harnesses are comfortable.

Birds will be in full song each morning and the rooks will be active in the rookeries.

APRIL

APRIL SHOWERS and warmer weather bring grass growth and leaves give a green haze to the hedgerows. The cattle will skip and gambol onto fresh pastures after their winter confinement in the yard. Ensure that they do not eat too much new grass and become 'blown'. The milk will take on a fresh colour and will be especially good for butter and cheese. The spring milk is also good for puppies and according to William Cobbett: 'Feed young puppies upon milk from the cow and they will never die with that ravaging disease called "the distemper".'

Be sure to set an even number of hen's eggs, and dampen the eggs of ducks to ensure many young. Do not pick wildflowers while a hen is sitting.

Cuckoos and swallows will confirm the arrival of spring.

MAY

PRIMROSES SHOULD BE PLACED in the cowshed on the 1st to keep witches and evil influences away. All the cattle, except the calves, will be spending their nights out. Pigs, too, can be let out from May until November, to make them fat. A start should be made to clean out the pig styes and the cowsheds.

Country cats and dogs can spend their days in the garden and freedom will make them fit. Caged birds can now have plenty of 'green stuff' to ensure good health. Egg numbers fall and broody hens should be taken from the nest boxes.

The sheep should be washed in readiness for shearing, but: 'Shear your sheep in May and you shear them all away.'

Hay making starts and the summer flowers begin to bloom.

JUNE

THIS IS THE TIME to get as much good hay as possible for winter fodder, so: 'Make hay while the sun shines.'

With the arrival of the elder flower it is also time to shear the sheep. Lambs should be dipped, wormed and protected against foot rot. If rot develops: 'They will limp and nod at you as if you are the parson.'

Despite the warmer weather any chicks or ducklings must be protected from getting cold if they are caught in showers or thunder storms.

Root crops grow faster and the cereal crops come onto ear; it is another good month for wildflowers.

JULY

THIS IS GENERALLY the hottest month and the pigs should be given a muddy wallow for comfort, cleanliness, and even for weather-forecasting:

> 'When pigs carry sticks, the clouds will play tricks,
> When they lie in the mud, no fears of a flood.'

Bryony and elder leaves should be used to keep flies away from horses. If cows get warble fly larvae, as swellings along their backs, turpentine should be poured into the 'maggots' breathing holes.

Ground ivy can be picked for eyewashes and eczema ointment, and raspberries may be dried to help with pregnancies later on. Even ragwort should be picked for use. When it grows in grass meadows it is a nuisance, but odd plants can be brewed like tea to treat the pads of dogs when they have been torn or bruised on hard ground.

AUGUST

IN A NORMAL YEAR this should be the main harvest month and it is also a time for harvest flowers, if the weedkillers have missed them.

After the passage of the combine harvester, hens should be put onto the stubble to clear up the wild oats. Hens that have moulted need plenty of protein to encourage the rapid growth of new feathers. Hen houses should be prepared for winter.

Milk yield will be at its lowest now and the cowshed should be treated for flies to prevent discomfort during milking.

Dogs will treat themselves if off-colour with couch grass and cats will look for cat-mint. Walnut leaves should be picked and dried for later use against worms.

SEPTEMBER

THIS IS THE LAST MONTH of summer when the harvest is usually completed and the nights begin to 'draw-in'.

The last of the thistles can be cut in the meadows and with the water table low, cattle drinking places can be cleaned out in streams and ponds. It is also a good time to clear unwanted weed from the garden pond.

The breeding ewes must be selected and pullets should be moved into clean henhouses, with increased food, to get them into lay.

Pigs will do well on the harvest stubble and in the orchard on fallen apples and late plums.

Swallows gather on telephone wires and flocks of starlings and lapwings are more evident.

OCTOBER

OCTOBER HERALDS the arrival of autumn. Sheds, hutches and kennels should be made water and draught proof, ready for the winter. Dogs and cats ought to be wormed and deloused, to keep the parasites picked up from their summer hunting, out of the house.

Corn, apples and straw must be stored in dry places and mangolds, turnips and swedes should be put into clamps.

The pigs can be moved into woodland where they will thrive on acorns, and hazel and beech mast. In lowland areas it will be 'tupping' time at the middle of the month.

Work horses must be watched carefully for digestive troubles when they begin to lie in at night and pullets must be well fed to prevent a moult.

Now is a good time for spreading manure:

'In October muck your field,
And your land its wealth will yield.'

49

NOVEMBER

THE LAST OF THE MANGOLDS should be lifted and land work will continue if heavy rain holds off. Horses and cattle can be left in the fields during the day until the ground becomes too wet. Carrots fed with winter rations will help put summer shines on winter coats and cool cattle sheds will encourage quicker growth.

Cockerels and Christmas turkeys should be putting on weight and mashed potatoes can be fed to geese. Nest boxes must contain plenty of clean straw to prevent dirty eggs. The pigs ought to be confined to their styes once more.

The 'louseberries' should be picked from spindle trees, as should rose hips for worm powder and the berries of mountain ash to aid foaling. Both garden and field will adopt the appearance of real winter.

DECEMBER

DECEMBER IS OFTEN a damp gloomy month. It is gloomy for some of the pigs too, as it is now time for them to be really 'cured', and: 'A hog is never good until it is on the dish'. Cobbett recommended December for making bacon, especially 'fat' bacon: 'Lean bacon is the most wasteful thing that any family can use. In short it is uneatable, except by drunkards who want something to stimulate their sickly appetite'.

Warmth and good food becomes essential for all stock, both in the house and in the yard. Sheep should be watched for liver fluke and worms.

December is also a good time to take out brambles from hedges which can get caught up in wool.

Index

51

Rooks, 46
Rose, wild, 14
Rosemary, 6, 18, 29
Rot, 17, 47
Rowan, 25, 29
Rue, 27, 29
Rutting, 15

Salt, 5, 17, 18, 25, 32
Salts: Epsom, 20; cattle, 31
Scab, 17, 18
Scour, ix, 14, 25, 33, 34, 38
Seaweed, 4
Shearing, 16
Sheep, 15-18, 22, 27, 47, 50;
    -dog, 2, 5; Jacob's, 12
Shock, 8
Skin, 7
Sloe wine, ix, 33
Slugs, 6
Snails, viii, 21, 22
Snakes, 34; bites, 34
Soap, 6, 7, 9, 14, 20
Soda, 6, 31; bicarbonate, 11;
    caustic, 8, 33
Sodium sulphate, 5
Sows, 12-15
Spanish fly, 31
Spindle tree, 2, 50
Squills, 31
Stable, ix
Stallion, 24-5
Starlings, 49
Stockholm tar, 26
Strawberry, 6
Suet, 17
Swallows, 40, 46, 49
Swede, 7, 49
Syrup of figs, 36

Tansy leaves, 2
Tartarised antimony, ix
Teeth, 8, 10, 24, 36
Thrush, 26
Thyme, 7

Ticks, 10, 14, 17
Tincture of benzoin, 6
Tincture of rhubarb, ix
Toad, 28
Toadmen, 28, 29
Tobacco, 14, 17, 28, 34
Train oil, 5
Treacle, 18, 35
'Trembles', 18
Tumours, 8
'Tupping', 49
Turkeys, 21, 50
Turnip, 7, 15, 26, 49
Turpentine, 8, 27, 33, 48
Twitch, 4

Udder, 33-4
Ulcers, 35
Urine, human, ix, 14, 17

Valentine's Day, St., 45
Valerian, 11, 40
Vaseline, 6
Vinegar, 6, 8, 14, 28, 31
Violets, 8
Vipers, 14, 34
Vitriol, 17

Walnut leaves, 3, 26, 48
Warble fly, 48
Warts, viii
Wasps, 6
Water mint, 7
Weasels, 34
Wheat, 25
Whey, 33
White chute, 25
Whitewash, 6
Wine: sloe, ix, 33; white, 18
Woodeford, James, 36
Worms, 2, 4, 10, 14, 18, 19, 28, 33,
    45, 48, 50; oil, 25; powder, 50

Yeast, 18
Yellow diarrhoea, 21